Nanami

Hiroto

⑪ Addition

1 There are 9 children on the bus.
If 4 more children get on the bus, how many children will there be altogether?

① Let's write a math expression.

Is the answer larger than 10?

Hiroto

② Let's think about how to calculate.

$9 + 4$

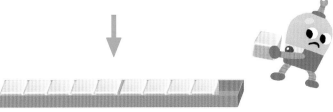

To make 10, 1 should be added to 9.

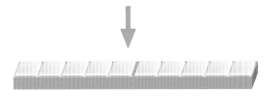

Decompose 4 into 1 and 3, then add 1 to 9.

10 and 3 make ⬚.

Math Sentence : $9 + 4 =$ ⬚   Answer : ⬚ children

How to add $9 + 4$

(1) To make 10, ⬚ should be added to 9.

(2) Decompose 4 into ⬚ and ⬚.

(3) Add 9 and ⬚ to make 10.

(4) 10 and ⬚ make ⬚.

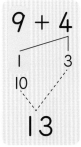

Let's talk about how to calculate 8 + 3.

To make 10, ...

Hiroto

(1) To make 10, ☐ should be added to 8.

(2) Decompose 3 into ☐ and ☐.

(3) Add 8 and ☐ to make 10.

(4) 10 and ☐ make ☐.

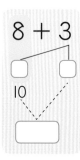

8 + 3

10

 Let's find the answers.

① 9 + 3   ② 9 + 2   ③ 8 + 4

④ 8 + 5   ⑤ 7 + 4   ⑥ 7 + 5

3 How many eggs are there altogether?

Let's think about how to calculate.

$3 + 9$

What should I do to make 10?

Instead of 9, I should decompose 3 ...

Yui

Daiki

2 Let's find the answers.

① $2 + 9$ ② $3 + 8$ ③ $4 + 9$

④ $4 + 7$ ⑤ $5 + 8$ ⑥ $4 + 8$

Activity

4 Let's think about how to calculate $8 + 6$.

How can I make 10?

Nanami

① What are the similarities and differences among the ideas of the following 3 children.

Hiroto's idea

(1) To make 10, 2 should be added to 8.
(2) Decompose 6 into 2 and 4.
(3) Add 8 and 2 to make 10.
(4) 10 and 4 make 14.

Yui's idea

(1) To make 10, 4 should be added to 6.
(2) Decompose 8 into 4 and 4.
(3) Add 6 and 4 to make 10.
(4) 10 and 4 make 14.

Nanami's idea

(1) Decompose 8 into 5 and 3, and 6 into 5 and 1.
(2) Add 5 and 5 to make 10.
(3) As for remainders, add 3 and 1 to make 4.
(4) 10 and 4 make 14.

② Let's talk about Nanami's idea by using blocks.

5 There were 5 doves. 6 doves flew in.
How many doves were there altogether?

 Let's find the answers.

① $9 + 8$	② $7 + 6$	③ $8 + 7$
④ $6 + 9$	⑤ $7 + 9$	⑥ $8 + 9$
⑦ $8 + 8$	⑧ $7 + 7$	⑨ $6 + 7$
⑩ $6 + 6$	⑪ $9 + 9$	⑫ $6 + 8$
⑬ $9 + 5$	⑭ $6 + 5$	⑮ $5 + 9$

 6 Let's make a math problem for $7 + 8$.

Addition Cards

Let's make addition cards and practice addition facts.

card

front 8 + 3

back | |

 Say the answer.

Lining up the cards in order

9 + 2	8 + 3	7 + 4	6 + 5	5 + 6	4 + 7
9 + 3	8 + 4	7 + 5	6 + 6	5 + 7	4 + 8
9 + 4	8 + 5	7 + 6	6 + 7	5 + 8	4 + 9
9 + 5	8 + 6	7 + 7	6 + 8	5 + 9	
9 + 6	8 + 7	7 + 8	6 + 9		
9 + 7	8 + 8	7 + 9			
9 + 8	8 + 9				
9 + 9					

Let's talk about what you noticed.

2 # Let's play a game.

The card with the answer of 15.

3 + 8   2 + 9

3 + 9

If the number being added increases by 1, the answer ...

Nanami

What you can do now

☐ Can do addition.

1 Let's find the answers.

① 9 + 6   ② 8 + 3   ③ 7 + 5   ④ 7 + 8

⑤ 3 + 9   ⑥ 8 + 8   ⑦ 5 + 7   ⑧ 5 + 9

⑨ 5 + 6   ⑩ 6 + 9   ⑪ 9 + 9   ⑫ 8 + 7

☐ Can make an addition expression and find the answer.

2 Let's write a math expression and find the answer.

① There are 8 pencils in the pencil case and 4 pencils on the desk. How many pencils are there altogether?

② I picked 9 persimmons yesterday and 7 persimmons today. How many persimmons did I pick altogether?

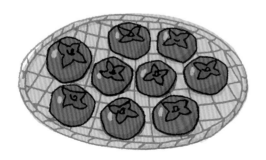 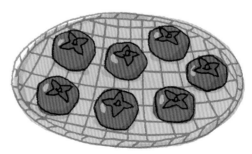

12 Subtraction

1 There were 12 acorns. I used 9 of them to make tops.

How many acorns were left?

What should I do to take away 9?

Yui

① Let's write a math expression.

② Let's think about how to calculate.

$$12 - 9$$

Decompose 12 into 10 and 2.

Subtract 9 from 10.

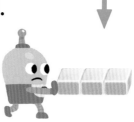

2 added to 1 is ☐.

Math Sentence : $12 - 9 =$ ☐ Answer : ☐ acorns

How to subtract $12 - 9$

(1) 9 cannot be subtracted from 2.

(2) Decompose 12 into 10 and 2.

(3) 10 minus 9 is ☐.

(4) ☐ added to ☐ is ☐.

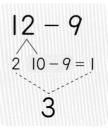

$$12 - 9$$
$$2 \quad 10 - 9 = 1$$
$$3$$

2 Let's talk about how to calculate $13 - 8$.

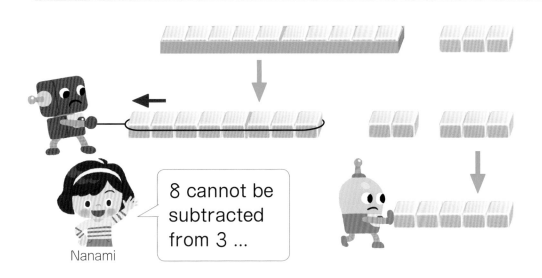

8 cannot be subtracted from 3 ...

Nanami

(1) 8 cannot be subtracted from 3.

(2) Decompose 13 into 10 and 3.

(3) 10 minus 8 is ☐.

(4) ☐ added to ☐ is ☐.

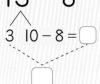

$13 - 8$

3 $10 - 8 = $ ☐

☐

 1 Let's find the answers.

① $16 - 9$ ② $11 - 9$ ③ $14 - 9$

④ $13 - 9$ ⑤ $14 - 8$ ⑥ $15 - 8$

⑦ $11 - 8$ ⑧ $13 - 7$ ⑨ $15 - 9$

3 There are 11 chocolates. If you eat 2 chocolates, how many chocolates will be left? Let's think about how to calculate.

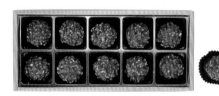

If I eat first the one outside the box ...

Hiroto

$11 - 2$

2 Let's find the answers.

① $12 - 3$ ② $11 - 3$ ③ $16 - 8$

④ $14 - 5$ ⑤ $17 - 8$ ⑥ $16 - 7$

Activity

4 Let's think about how to calculate $14 - 6$.

Take 6 from 10 in the box ...

How about taking 4 first?

Daiki

Yui

14

① What are the similarities and differences between Daiki's and Yui's ideas.

 Daiki's idea

(1) 6 cannot be subtracted from 4.

(2) Decompose 14 into 10 and 4.

(3) 10 minus 6 is 4.

(4) 4 added to 4 is 8.

$$14 - 6$$
4 10 − 6 = 4
8

 Yui's idea

(1) 6 cannot be subtracted from 4.

(2) Decompose 6 into 4 and 2.

(3) 14 minus 4 is 10.

(4) 10 minus 2 is 8.

$$14 - 6$$
4 2
14 − 4 = 10
10 − 2 = 8

② Let's talk about Yui's idea by using blocks.

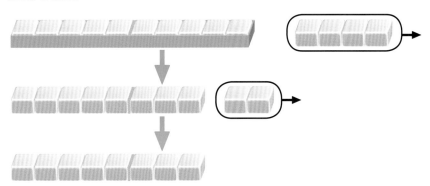

3 ▶ Let's find the answers.

① 11 − 5 ② 12 − 6 ③ 13 − 5

④ 14 − 7 ⑤ 17 − 9 ⑥ 13 − 4

⑦ 11 − 3 ⑧ 18 − 9 ⑨ 13 − 6

⑩ 12 − 3 ⑪ 15 − 6 ⑫ 11 − 4

⑬ 11 − 6 ⑭ 15 − 8 ⑮ 13 − 7

Want to try **Various subtractions**

5 Let's subtract each of the surrounding numbers from the number in the center.

①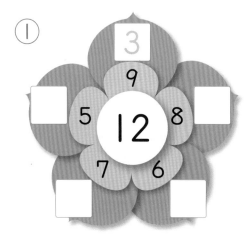

②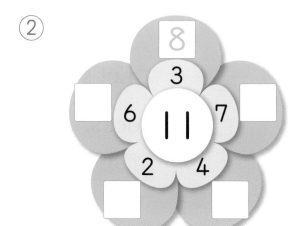

4 Let's find the answers.

① $12 - 5$ ② $15 - 9$ ③ $14 - 8$

④ $17 - 8$ ⑤ $16 - 9$ ⑥ $15 - 7$

⑦ $11 - 7$ ⑧ $13 - 8$ ⑨ $12 - 4$

 5 Daiki picked up 9 leaves and Nanami picked up 13 leaves.

Who picked up more leaves and by how many?

Daiki

Nanami

Want to represent

6 Let's make a math problem for $12 - 5$.

Subtraction Cards

card

front 14−8

back 6

Let's make subtraction cards and practice subtraction facts.

 Say the answer.

6

Lining up the cards in order

11 − 2	12 − 3	13 − 4	14 − 5	15 − 6	16 − 7
11 − 3	12 − 4	13 − 5	14 − 6	15 − 7	16 − 8
11 − 4	12 − 5	13 − 6	14 − 7	15 − 8	16 − 9
11 − 5	12 − 6	13 − 7	14 − 8	15 − 9	
11 − 6	12 − 7	13 − 8	14 − 9		
11 − 7	12 − 8	13 − 9			
11 − 8	12 − 9				
11 − 9					

Let's talk about what you noticed.

Let's play a game.

The card with the answer of 5.

Picking up the card

Matching cards

$$17 - 8 \qquad 18 - 9$$

$$17 - 9$$

If the number being subtracted increases by 1, the answer ...

Nanami

Comparing answers

Want to know

1 How many monkeys are there altogether?

Want to try

1 There were 16 apples.

The elephant ate 7 apples.

How many apples were

left?

Want to know

2 There are 8 penguins on an island. If 3 more penguins come in, how many penguins will there be altogether?

Want to try

 There are tigers and lions. Which is more and by how many?

What you can do now

Can do subtraction.

1 Let's find the answers.

① $17 - 9$ ② $15 - 7$ ③ $11 - 4$

④ $13 - 6$ ⑤ $12 - 7$ ⑥ $11 - 5$

⑦ $11 - 8$ ⑧ $12 - 8$ ⑨ $14 - 6$

⑩ $18 - 9$ ⑪ $12 - 4$ ⑫ $16 - 8$

Can make a subtraction expression and find the answer.

2 Let's write a math expression and find the answer.

① An apple tree had 14 fruits. You picked 7 fruits.

How many fruits were left?

② Which group is bigger and by how many?

13 Comparing sizes

Want to compare

1 Which is longer?
Let's think about how to compare the lengths.

① Pencils

② Strings

Ⓐ Ⓑ

③ Postcard

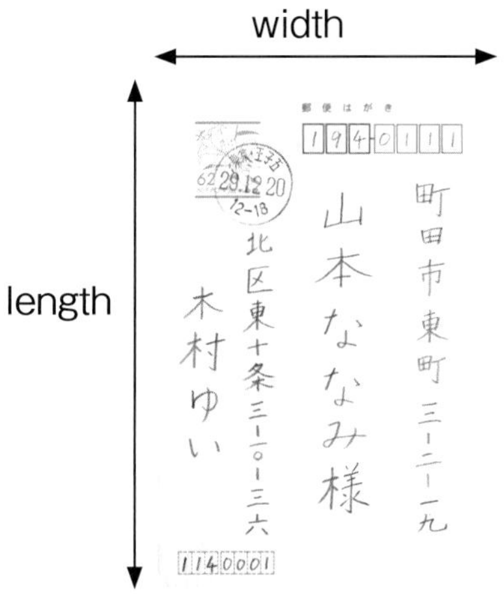

width

length

④ Box

width

length

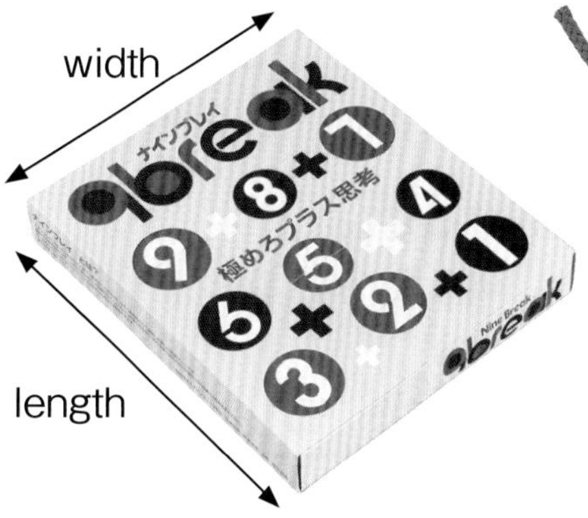

① Pencils   ② Strings   ③ Postcard   ④ Box

Fold and
overlap ...

Yui

We cannot fold
a box, ...

Nanami

Align their
edges ...

Hiroto

Align their
edges and
stretch ...

Daiki

2 Let's compare lengths by using a tape.

We can measure lengths such as a circumference by using a tape.

Can this desk go through the doorway?

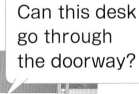

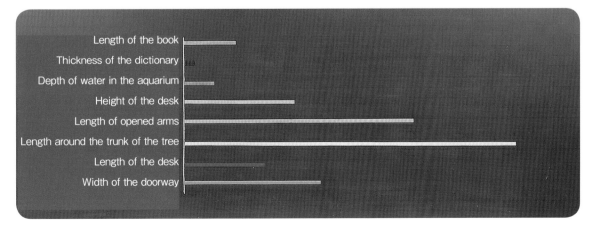

Length of the book

Thickness of the dictionary

Depth of water in the aquarium

Height of the desk

Length of opened arms

Length around the trunk of the tree

Length of the desk

Width of the doorway

3 Let's represent the width of the desk with a number.

6 units of the length between the tips of my thumb and forefinger.

4 units of the length of this pencil.

 Let's examine the following lengths.

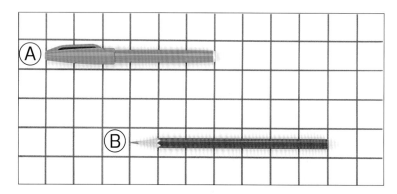

① How many grids represent the length of Ⓐ and Ⓑ?

② Which is longer and by how many grids?

1 Which bottle holds more juice? Let's think about how to compare the amounts.

I cannot compare by just looking at the bottles.

What should we use to compare them?

Hiroto

Nanami

① Let's talk about how to compare the amounts as shown below.

How to compare the amounts (1)

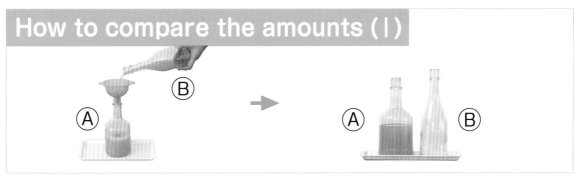

How to compare the amounts (2)

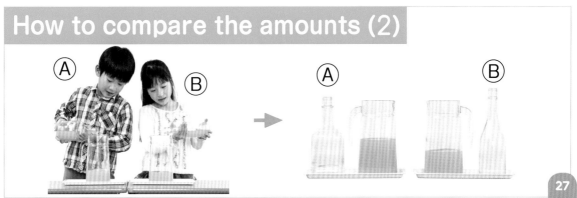

2 How much more juice is there in bottle Ⓐ than bottle Ⓑ?

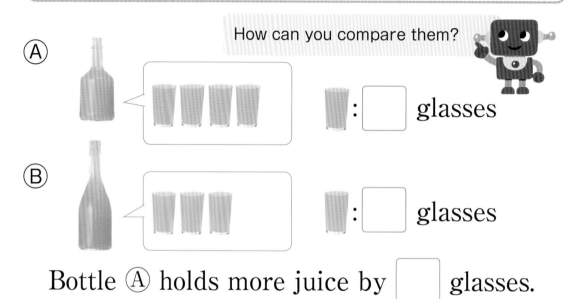

How can you compare them?

Ⓐ 🍶 : ☐ glasses

Ⓑ 🍶 : ☐ glasses

Bottle Ⓐ holds more juice by ☐ glasses.

Amount of water

1 Which container can hold more water and by how much?

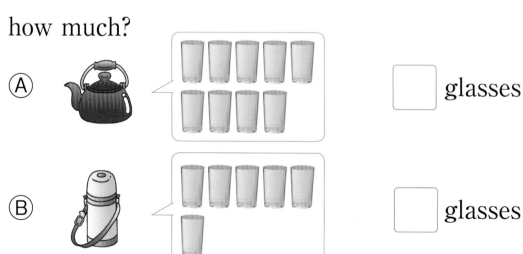

Ⓐ ☐ glasses

Ⓑ ☐ glasses

☐ can hold more water by ☐ glasses.

2 Let's arrange the following containers Ⓐ, Ⓑ, and Ⓒ according to amount in descending order.

3 Which box is larger?

①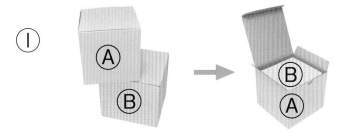

②

Want to compare

1 Which is larger?

①

②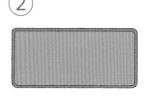

Want to try

▶ 1 Let's play an encampment game.

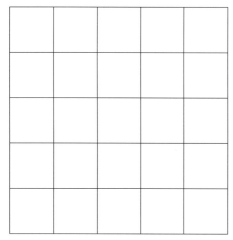

How to play

❶ Let's play the rock-paper-scissors. If you win, you can paint one grid in your color.

❷ The one who paints the largest area wins.

What you can do now

Can compare lengths.

1 Which train is the longest?

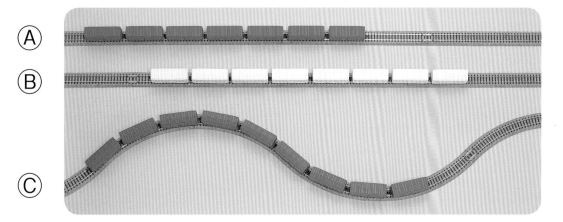

Can compare amounts.

2 Which one contains more? Let's talk about how
to compare the water bottles.

Can compare areas.

3 Which is larger?

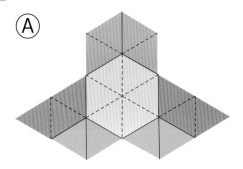
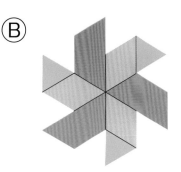

Let's look at the scenes in the classroom at the 4th period and noon break. Can you find any difference between them?

Want to explore   In the classroom

The scenes in the classroom at the **4**th period and noon break are shown below.

4th period

Noon break

1 Let's find a lot of differences between the scenes at the 4th period and noon break, then make a presentation about them.

2 Let's represent the differences between the scenes at the 4th period and noon break in addition and subtraction expressions, then make math stories.

Have you made a lot of stories?

Think in groups

3 Let's make a presentation to the group about the stories you made. Then, decide the favorite math expression and story.

First of all, the order of presentations should be decided. Then let's give a presentation in order.

Think as a class

4 Let's make a presentation of the favorite story in your group.

Which story was the hardest for you to understand?

Nanami

01102

14 Numbers Larger than 20

Want to know

1 How many ▢ are there?

Yui

Daiki

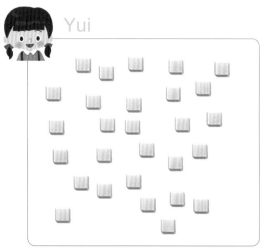

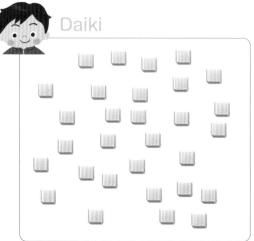

① Let's count each number of blocks by lining them up to see the number easily.

② What is the number of Yui's 🧽 ?

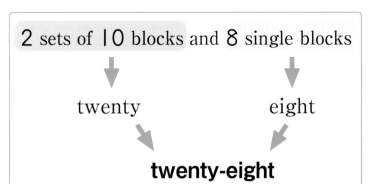

sets of 10 blocks	single blocks

Tens place	Ones place
2	8

For 28,

the number in the **tens place** is ☐ and

the number in the **ones place** is ☐ .

2	8

③ What is the number of Daiki's 🧽 ?

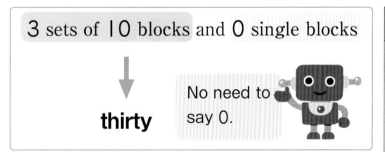

sets of 10 blocks	single blocks

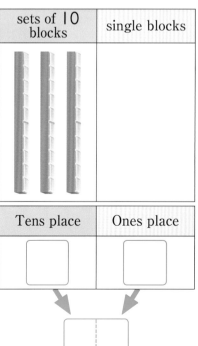

Tens place	Ones place
☐	☐

For 30,

the number in the tens place

is ☐ and the number in the

ones place is ☐ .

2 Let's write the following numbers.

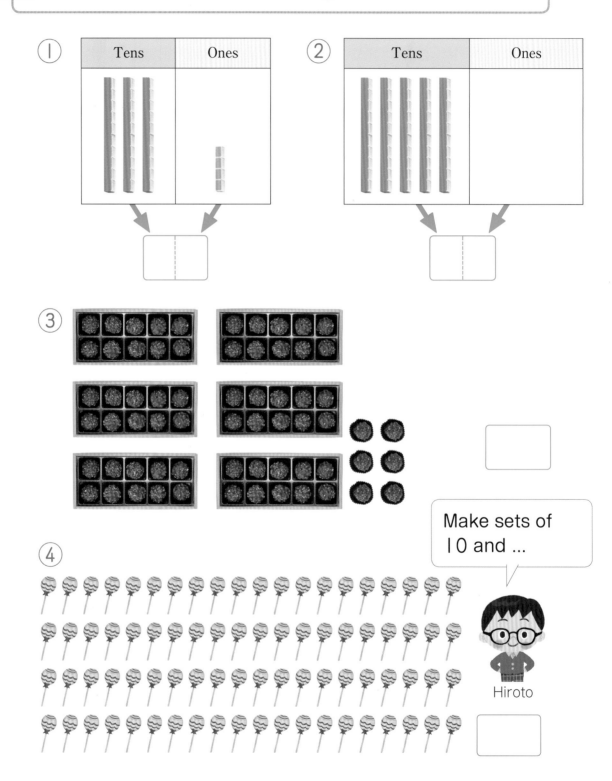

Make sets of
10 and ...

Hiroto

3 Let's represent 46 by using .

Forty-six

46

Since there are 4 sets of 10 blocks and 6 single blocks, ...

1 Let's fill in each ☐ with a number.

① 8 sets of 10 and 2 ones make ☐.

② 9 sets of 10 make ☐.

③ 74 is ☐ sets of 10 and ☐ ones.

④ 60 is ☐ sets of 10.

2 Let's fill in each ☐ with a number.

① The number that has 9 in the tens place and 5 in the ones place is ☐.

② For 80, the number in the tens place is ☐ and the number in the ones place is ☐.

How many fish are there?

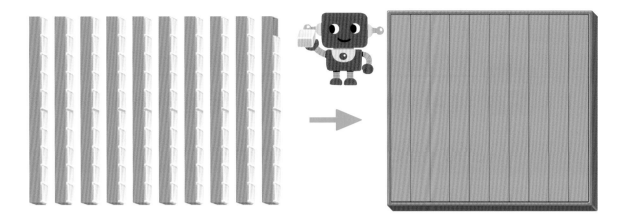

|0 sets of |0 make **one hundred**. It is written as |00. |00 is the number that is | larger than 99.

Want to try

3 Let's write the following numbers.

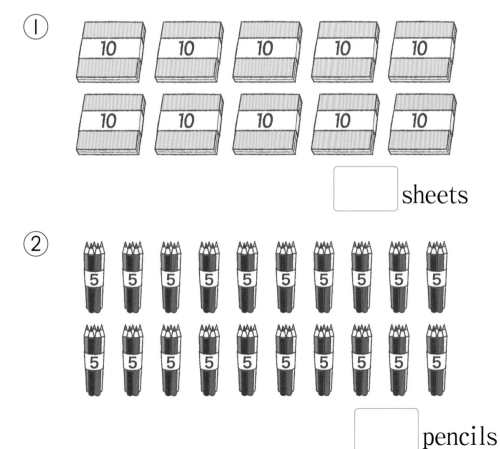

① sheets

② pencils

5 Let's make number cards from 0 to 100 and place them in order.

0	1	2	3	4	5	6	7	8	9
10	11	12	13	14	15	16	17	18	19
20	21	22	23	24	25	26	27	28	29
30	31	32	33	34	35	36	37	38	39
40									
50									59
	61				65				
70									
	81								89
90						96		98	
100									

① Let's find numbers that have **7** in the ones place.

② Let's find numbers that have **8** in the tens place.

0   10   20   30   40   50   6

4 Which number is larger?

① 47 58   ② 78 75

③ 89 98   ④ 61 59

5 Let's fill in each ☐ with a number.

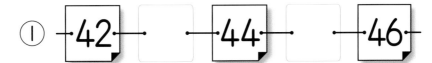

① 42 ☐ 44 ☐ 46

② ☐ ☐ 98 99 ☐

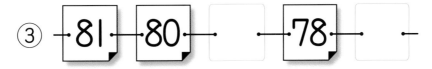

③ 81 80 ☐ 78 ☐

6 Let's fill in each ☐ with a number.

① The number that is 3 larger than 97 is ☐ .

② The number that is 10 smaller than 100 is ☐ .

70   80   90   100   110   120

Want to know

1 How many pencils are there?

100 and 12 make

112. It is read as

one hundred twelve.

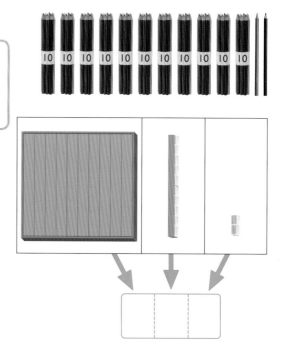

Want to try

1 Let's write the following numbers.

① ②

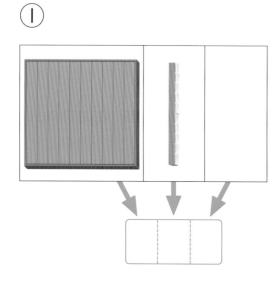 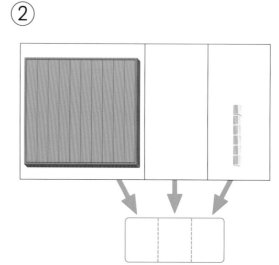

1 There are 20 sheets of red origami paper and 30 sheets of blue origami paper.
How many sheets are there altogether?

① Let's write a math expression.

② Let's think about how to calculate.

Answer : ____ sheets

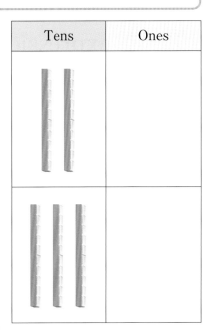

Tens	Ones

Thinking of sets of 10 ...

Daiki

 Let's find the answers.

① 40 + 30   ② 10 + 80

③ 20 + 10   ④ 30 + 70

2 There are 23 crayons. If you get 6 more crayons, how many crayons will you have altogether?

① Let's write a math expression.

② Let's think about how to calculate.

Answer : [   ] crayons

Tens	Ones
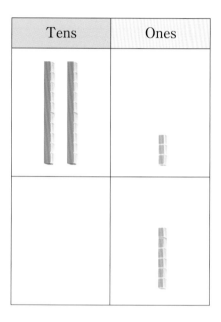	

 2 There are 34 children and 3 adults. How many people are there altogether?

 3 Let's find the answers.

① 42 + 1   ② 25 + 4   ③ 36 + 2

④ 70 + 5   ⑤ 4 + 52   ⑥ 6 + 33

⑦ 3 + 21   ⑧ 9 + 60   ⑨ 5 + 83

3 There were 50 strawberries.
20 strawberries were eaten. How many strawberries were left?

① Let's write a math expression.

② Let's think about how to calculate.

Nanami

Also for subtraction, thinking of sets of 10 ...

Tens	Ones

Answer : ☐ strawberries

4 There are 100 sheets of origami paper. If you use 50 sheets, how many sheets will be left?

5 Let's find the answers.

① 40 − 20   ② 90 − 30   ③ 100 − 40

4 There are 38 red birds and 5 white birds. What is the difference between the numbers of birds?

① Let's write a math expression.

② Let's think about how to calculate.

Answer : ___ birds

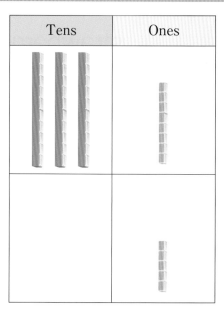

Tens	Ones

6 There are 24 cakes. If you eat 4 cakes, how many cakes will be left?

7 Let's find the answers.

① 48 − 3   ② 67 − 5   ③ 98 − 7

④ 26 − 2   ⑤ 37 − 7   ⑥ 55 − 5

⑦ 89 − 9   ⑧ 76 − 6   ⑨ 68 − 8

What you can do now

☐ Can count large numbers.

1 Let's write the following numbers.

① Colored pencils

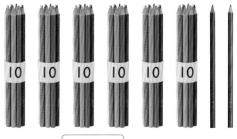

[   ] pencils

② Candies

[   ] candies

☐ Understanding the structure of large numbers.

2 Let's fill in each [ ] with a number.

① 9 sets of 10 and 8 ones make [   ].

② [   ] sets of 10 and [   ] ones make 67.

③ [   ] should be added to 96 to make 100.

④ The number that is 10 smaller than 120 is

[   ].

☐ Can calculate large numbers.

3 Let's find the answers.

① 50 + 30   ② 32 + 6   ③ 8 + 41

④ 70 − 20   ⑤ 86 − 4   ⑥ 58 − 8

15 Time (2)

1 Let's look at the above pictures and talk about them.

The short hand shows the hours and the long hand shows the minutes.

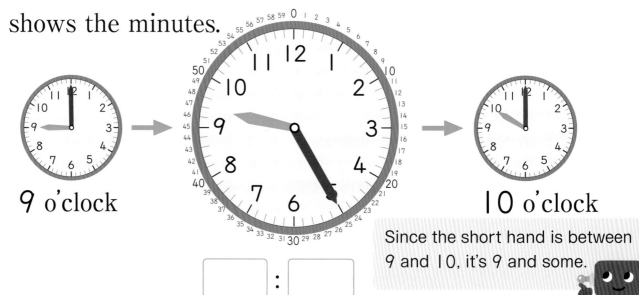

9 o'clock

10 o'clock

Since the short hand is between 9 and 10, it's 9 and some.

☐ : ☐

48

How many minutes does one scale represent?

Yui

Let's think by using the clock on page 84.

10:30 can also be read as half past ten.

Hiroto

It's also read as three to two.

Daiki

It's a little past three.

Nanami

2 Let's show the time by moving the hands of the clock.

① 1 : 23

② 5 : 45

③ 11 : 04

First count 5, 10, 15, and 20, then move to 21, 22, and 23.

Hiroto

1 Let's connect each time to the clock that represents it.

| 2 : 15 | 5 : 55 | 9 : 09 | 11 : 30 |

2 Let's talk about what time you got up.

I got up at 6:30 and washed my face.

Yui

I got up at 7:10. I overslept.

Daiki

16 Addition or Subtraction? Let's think by using diagrams.

Want to know How many children altogether?

1 The children are hiking. Aya is the 6th child from the front. There are 3 children behind her.

How many children are there altogether?

6th

front ⭕⭕⭕⭕⭕🔵⭕⭕⭕ back

☐ children ☐ children

Math Expression : ☐

Answer : ☐ children

1 There are 10 children looking at the cows. Kenta is the 4th children from the left.

How many children are there to the right of him?

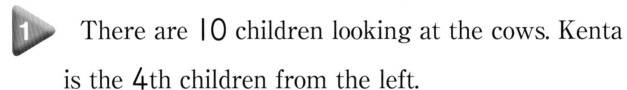

4th
↓
left ○○○ ● ○○○○○○ right

[ ] children   [ ] children

Math Expression : [ ]

Answer : [ ] children

How many altogether?

2 There are 7 children and each one has 1 ball. There are 4 balls left in the basket. How many balls are there altogether?

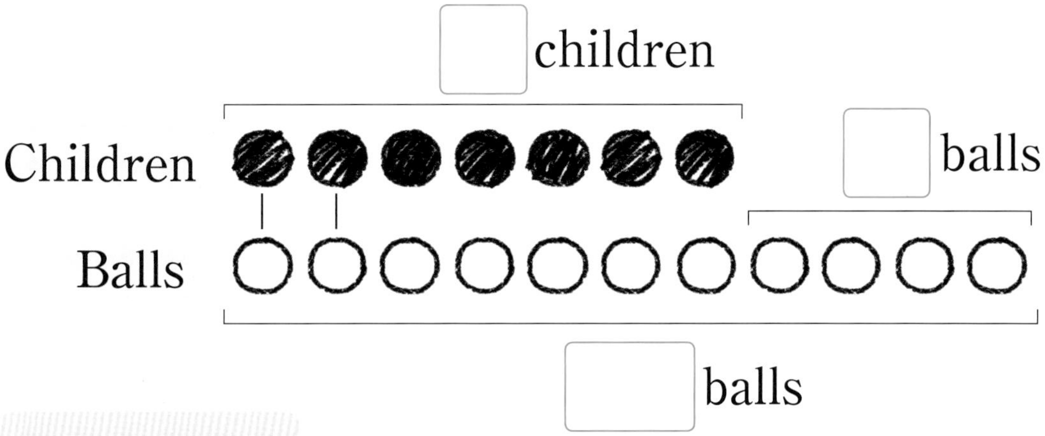

☐ children

Children ●●●●●●● ☐ balls

Balls ○○○○○○○○○○○

☐ balls

Let's connect
● and ○ with
lines.

In the diagram,
what represents
the answer?

Yui

Math Expression : ☐

Answer : ☐ balls

Want to try How many left?

 There are 9 cakes.

If 5 children get 1 cake each, how many cakes will be left?

◻ cakes

Cakes ● ● ● ● ● ● ● ● ●

Children ○ ○ ○ ○ ○

◻ children

Let's think by connecting ● and ○ with lines.

Math Expression : ◻

Answer : ◻ cakes

54

3 There are 6 apples. Oranges are 4 more than apples.

How many oranges are there?

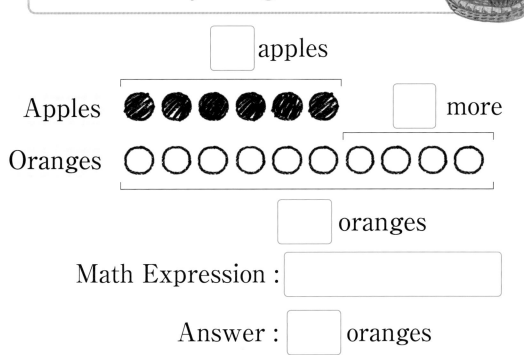

☐ apples

Apples ●●●●●● ☐ more

Oranges ○○○○○○○○○○

☐ oranges

Math Expression : ☐

Answer : ☐ oranges

Want to try

3 There are 7 red goldfish in the aquarium.

Black goldfish are 5 more than red goldfish.

How many black goldfish are there?

Red ○○○○○○○○○○○○

Black ○○○○○○○○○○○○

4 Saki picked up 10 acorns. Her sister picked up 3 fewer acorns than Saki. How many acorns did her sister pick up?

☐ acorns

Saki ⭕⭕⭕⭕⭕⭕⭕⭕⭕⭕

Sister ⭕⭕⭕⭕⭕⭕⭕ ○ ○ ○

☐ acorns   ☐ fewer

Math Expression : ☐

Answer : ☐ acorns

Want to try

4 In the ball-toss game, the Red team got 15 balls into the basket. The White team got 6 fewer balls than the Red team.

How many balls did the White team get?

Red ○○○○○○○○○○○○○○○

White ○○○○○○○○○○○○○○○

1 Let's share the strawberries equally between 2 children.

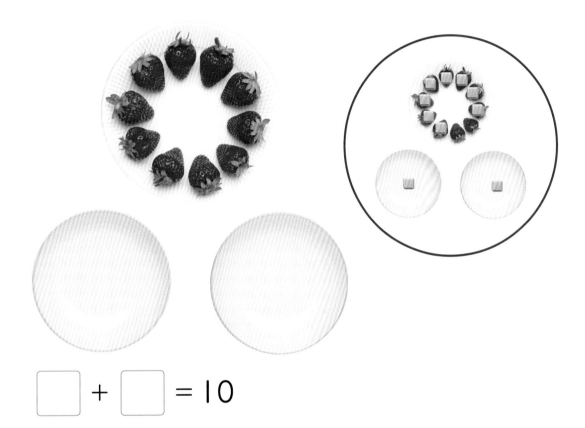

☐ + ☐ = 10

1 Let's share the candies equally among 3 children.

☐ + ☐ + ☐ = 18

Reflect Connect

Problem
Let's find 12.

Calendar
A year has 12 months.

There are 12 animal signs.

The twelve signs of the Japanese zodiac

Does 12 have any relationships with time, months, or years?

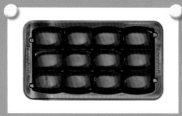  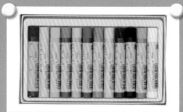

There are 12 pieces in a box.
It can be called 1 dozen.

12 is a number that is used a lot in our surroundings.

Let's find 4 from your surroundings.

I found 4 a lot in our classroom.

There are 4 outlets.

Hiroto

A chair has 4 legs.

Daiki

Let's examine how to arrange 12 pieces of ○.

Chocolates

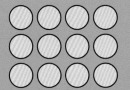

3 rows of 4 pieces

4 columns of 3 pieces

Math Sentence
$4 + 4 + 4 = 12$

Math Sentence
$3 + 3 + 3 + 3 = 12$

Pencils

2 rows of 6 pencils

6 columns of 2 pencils

Math Sentence
$6 + 6 = 12$

Math Sentence
$2 + 2 + 2 + 2 + 2 + 2 = 12$

Let's arrange 12 pieces of ○ and represent it in a math sentence.

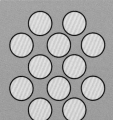

Math Sentence
$2 + 3 + 2 + 3 + 2 = 12$

Looking diagonally ...
$2 + 4 + 4 + 2 = 12$
$2 + 8 + 2 = 12$
$6 + 6 = 12$

When you change the way to arrange them, it can be represented in other math sentences.
As for other numbers, can you find various ways to arrange them?

There are 4 doors.

Nanami

We have 4 classes a day.

Yui

Want to connect

I want to examine about larger numbers.

Hiroto

Want to know

1   Hiroto's class has collected plastic bottle caps.

Let's look at the following blackboard and think about how many caps Hiroto brought this week.

On Monday, I brought 3 caps.

On Tuesday, 2 more caps than Monday.

On Wednesday, 3 fewer caps than Tuesday.

On Thursday, 4 more caps than Wednesday.

On Friday, the same as Thursday.

Daiki

It is difficult to see the number on each day as it is.

Can I organize the number of caps he brought on each day?

Yui

①   Let's write the number of caps brought on each day.

On Monday   ☐ caps

On Tuesday   since   ☐ + 2 = ☐ , ☐ caps

On Wednesday   since   ☐ − 3 = ☐ , ☐ caps

On Thursday   since   ☐ + 4 = ☐ , ☐ caps

On Friday   ☐ caps

②   Let's color the number of caps on each day.

How many caps were brought from Monday to Friday altogether?

Hiroto

Monday	Tuesday	Wednesday	Thursday	Friday

③   What is the difference between the largest number and the smallest number?

18 Shapes (2)

Want to represent Arranging colored pieces

1 Let's arrange the colored pieces to make different shapes.

You can use colored pieces on page 85.

2

Let's arrange 4 blue pieces  to make shapes.

① Let's make shape Ⓐ.

Ⓐ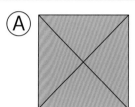

② From shape Ⓐ, let's make shapes Ⓑ, Ⓒ, and Ⓓ.

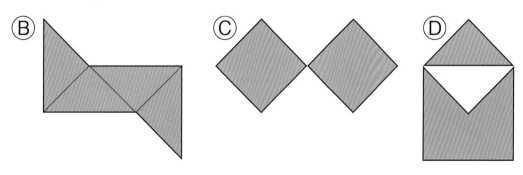

Ⓑ   Ⓒ   Ⓓ

1 Let's move one piece to make different shapes.

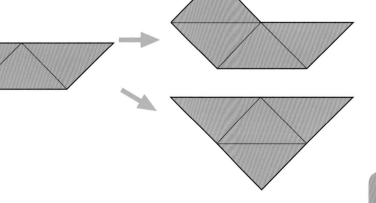

3 Let's arrange the sticks to make different shapes.

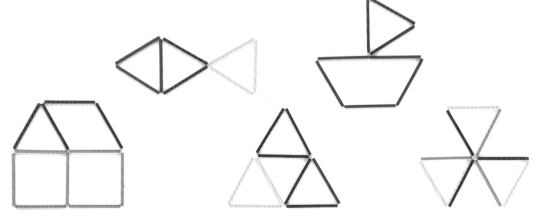

Want to confirm

2 Let's connect the dots to make different shapes.

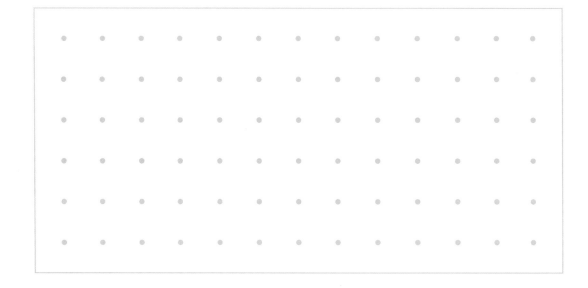

19 Summary of 1st Grade

Numbers Larger than 20

1 How many strawberries are there?

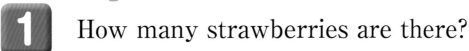

Numbers Larger than 20

2 Let's fill in each ☐ with a number.

① 7 sets of 10 and 6 ones make ☐ .

② 59 is ☐ sets of 10 and ☐ ones.

③ 70 is ☐ sets of 10.

④ The number that is 1 smaller than 100 is ☐ .

⑤ The number that is 1 larger than 119 is ☐ .

3 Which number is larger?

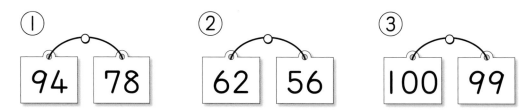

① 94 78 ② 62 56 ③ 100 99

4 Let's fill in each ☐ with a number.

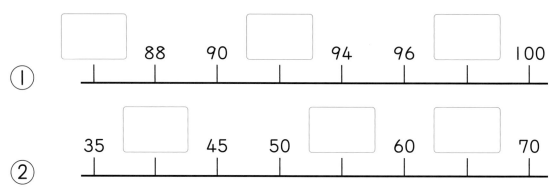

① [ ] 88 90 [ ] 94 96 [ ] 100

② 35 [ ] 45 50 [ ] 60 [ ] 70

5 There are 83 stickers. 10 stickers are put on each page.

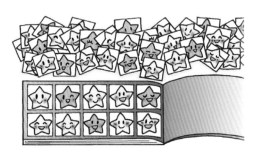

① How many pages are filled with 10 stickers?

② How many stickers are left?

6 Let's connect the dots from number 50 to 100.

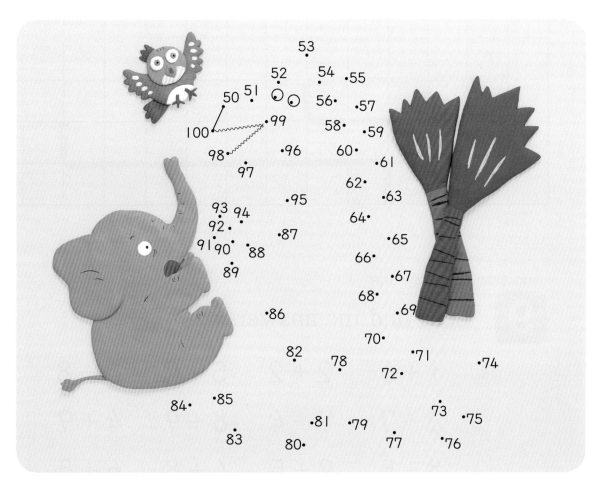

Time

7 What time is it?

①  ②  ③

Let's order the strings from the longest.

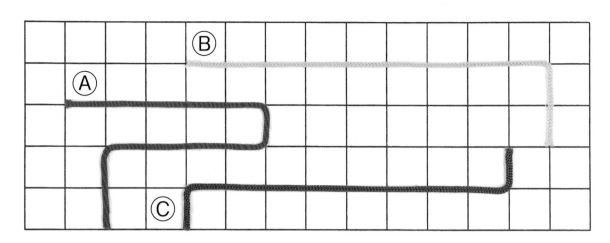

Addition, Subtraction

 Let's find the answers.

①   1 + 6   2 + 2   5 + 0   0 + 8

     3 + 7   7 + 4   8 + 9   4 + 9

     8 + 5   9 + 5   4 + 8   6 + 5

②   8 − 1   9 − 7   6 − 6   5 − 0

     10 − 8   11 − 3   12 − 4   14 − 9

     13 − 8   16 − 9   14 − 5   17 − 8

③   20 + 70   17 + 2   6 + 32   3 + 40

     70 − 30   65 − 2   47 − 7   90 − 90

10 Let's make math problems for the following math expressions.

① 8 + 4   ② 12 − 7

11 Akemi ate 7 cookies and her sister ate 6 cookies.

How many cookies did they eat altogether?

12 There were 12 children on the bus. At the bus stop, 6 children got off and 3 children got on.

How many children are there on the bus now?

Computational thinking

01103

Let's try moving Robo as you want.

Let's instruct Robo to move.

You can give 4 kinds of instructions to Robo.

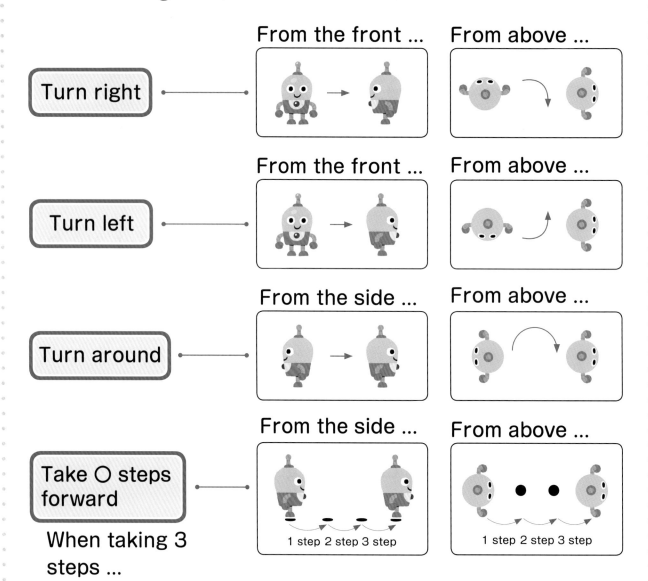

When taking 3 steps ...

1 If you give the following instructions to Robo, which direction will he face, Ⓐ or Ⓑ?

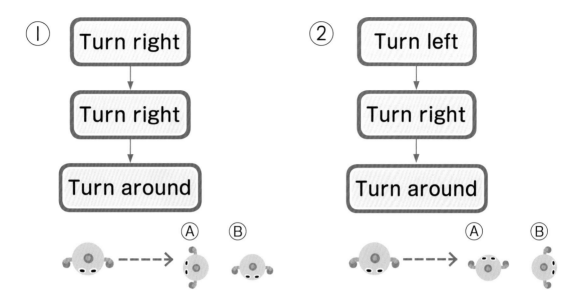

① Turn right → Turn right → Turn around

② Turn left → Turn right → Turn around

2 Let's instruct Robo to go to the place where there is a dog which failed to escape from the fire.

When Robo reaches the dog, he can help it.

Utilize math for our life

Let's make presents for the new first grade students.

I want to do something for the new students.

How about making a "math quiz about our school" and present the quiz to them.

The quiz will tell them many things about our school.

I hope they can find the answer through learning.

◯ I. Addition Quiz

Nanami

Our group will make an "addition quiz."

Addition Quiz

| There are 3 teachers in the first grade. There are 4 teachers in the second grade. How many teachers are there altogether? | First grade teachers Second grade teachers  Math expression : Answer : |

 Let's try making various addition quizzes.

Addition Story

"If you add a number of ◯◯ and a number of △△, what number will they be altogether?"

"If you add the number of △△ to the number of ◯◯, what number will it become?"

How many steps are there from the first floor to the landing and from the landing to the second floor altogether?

Daiki

2. Subtraction Quiz

Hiroto

Our group will make a "subtraction quiz."

Subtraction Story

"How many more (fewer) ○○ are there than △△?"
"There are □ ○○. If you use △ ○○, how many ○○ will be left?"

Subtraction Quiz

There are 15 children whose birthdays are from April to September. There are 9 children whose birthdays are from October to March. Which group has more children and by how many?	From April to September  From October to March  Math expression : Answer :

Let's try examining in your class.

? Let's try making various subtraction quizzes.

3. Shape Quiz

Yui

Our group will make a "shape quiz."

Shape Quiz

What is the shape of ○ found in the school? Let's try drawing it.	Example : unicycle

 Let's try making various shape quizzes.

4. Clock Quiz

Daiki

Our group will make a "clock quiz."

Clock Quiz

The clock on the right shows the time our school begins. What time is it?	

Want to make much more Let's also make a "size comparison quiz."

Utilize math for our life

What I can do now

1. Toward learning competency

	😊 Strongly agree	🙂 Agree	🙁 Don't agree
① It was fun making a quiz.			
② The learning contents were helpful.			

2. Thinking, deciding, and representing competency

	😊 Definitely did	🙂 So so	🙁 I didn't
① I was able to discover a quiz in which mathematics is used.			
② I was able to confirm whether I can solve the quiz.			
③ I was able to represent quizzes with words, pictures, and figures.			

3. What I know and can do

	😊 Definitely did	🙂 So so	🙁 I didn't
① I was able to make a better quiz.			
② I was able to solve the quiz I made.			

4. Encouragement for myself

	😊 Strongly agree
① I think that I'm doing my best.	

Give yourself a compliment since you have worked so hard.

Let's try to work out what you were not able to accomplish and keep doing your best on what you were able to fulfill. Then let's make one more quiz.

Supplementary Problems

⑪ Addition

pp.2 ~ 10

1 It is shown below how to calculate 8 + 5. Let's fill in each ☐ with a number.

(1) To make 10, ☐ should be added to 8.

(2) Decompose 5 into ☐ and 3.

(3) Add 8 and ☐ to make 10.

(4) 10 and ☐ make ☐.

2 Let's find the answers.

① 9 + 4 ② 9 + 5

③ 8 + 3 ④ 8 + 4

⑤ 7 + 4 ⑥ 6 + 5

3 Airi has 9 sheets of colored paper and her sister has 3 sheets.

How many sheets do they have altogether?

4 It is shown below how to calculate 5 + 7.

Let's fill in each ☐ with a number.

(1) To make 10, ☐ should be added to 7.

(2) Decompose 5 into ☐ and 2.

(3) Add ☐ and 7 to make 10.

(4) ☐ and 10 make ☐.

5 Let's find the answers.

① $3 + 8$ ② $4 + 9$

③ $4 + 8$ ④ $5 + 8$

⑤ $4 + 7$ ⑥ $5 + 6$

6 I ate 5 strawberries yesterday and 9 strawberries today.

How many strawberries did I eat altogether?

7 Let's find the answers.

① $8 + 6$ ② $9 + 7$

③ $6 + 8$ ④ $7 + 9$

⑤ $7 + 7$ ⑥ $9 + 8$

8 There are 7 doves. If 6 doves fly in, how many doves will there be altogether?

⑫ Subtraction
pp.11 ~ 22

1 It is shown below how to calculate $13 - 8$. Let's fill in each ☐ with a number.

(1) 8 cannot be subtracted from 3.

(2) Decompose 13 into ☐ and 3.

(3) ☐ minus 8 is 2.

(4) ☐ added to 2 is ☐.

2 Let's find the answers.

① $13 - 9$ ② $11 - 9$

③ $14 - 9$ ④ $12 - 8$

⑤ $14 - 8$ ⑥ $11 - 7$

3   There were 12 candies. I ate 9 candies. How many candies were left?

4   It is shown below how to calculate $12 - 4$. Let's fill in each ☐ with a number.
(1) 4 cannot be subtracted from 2.
(2) Decompose 12 into ☐ and 2.
(3) ☐ minus 4 is 6.
(4) ☐ added to 6 is ☐.

5   Let's find the answers.
① $13 - 5$ ② $15 - 6$
③ $16 - 7$ ④ $14 - 6$
⑤ $17 - 9$ ⑥ $13 - 4$

6   There were 15 children in the park. After a while, 7 children went home.
　How many children are there in the park?

7   By folding origami paper, Yuka made 5 cranes and Karin made 14 cranes. Who made more cranes and by how many?

8   There are 16 sheets of colored paper and their colors are red and blue. 9 sheets of them are red paper.
　How many sheets of blue paper are there?

13 Comparing Sizes

pp.23 ~ 31

1 Which is longer?

① Ⓐ

Ⓑ

② Ⓒ

Ⓓ

2 Which is larger?

Ⓐ

Ⓑ

3 Which is larger?

① Ⓐ

Ⓑ

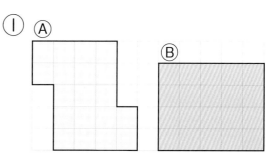

② Ⓒ

Ⓓ

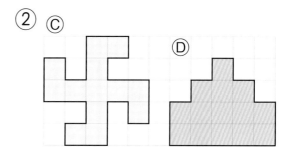

4 Which one contains more?

① Ⓐ   Ⓑ

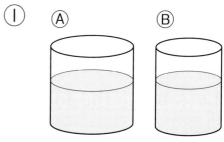

② Ⓒ

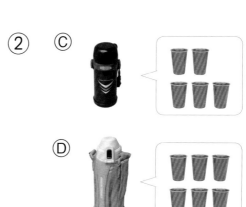

Ⓓ

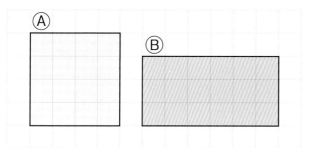

⑭ Numbers Larger than 20

pp.34 ~ 47

1 Let's write the following numbers.

①

②

2 Let's write the following numbers.

① The number that has 4 in the tens place and 6 in the ones place.

② The number that has 7 in the tens place and 9 in the ones place.

③ The number that has 5 in the tens place and 0 in the ones place.

3 Let's fill in each ▢ with a number.

① 6 sets of 10 and 3 ones make ▢.

② 3 sets of 10 and 9 ones make ▢.

③ 8 sets of 10 make ▢.

④ 92 is ▢ sets of 10 and ▢ ones.

⑤ 38 is ▢ sets of 10 and ▢ ones.

⑥ 70 is ▢ sets of 10.

⑦ 10 sets of 10 make ▢.

⑧ 100 is 1 larger than ▢.

④ Let's fill in each ☐ with a number.

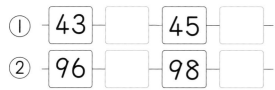

① 43 ☐ 45 ☐ ☐

② 96 ☐ 98 ☐

③ The number that is 5 larger than 55 is ☐.

④ The number that is 10 smaller than 87 is ☐.

⑤ The number that is 2 larger than 98 is ☐.

⑥ The number that is 3 smaller than 100 is ☐.

⑤ Let's say the following numbers in ascending order.

① 87, 78, 69

② 77, 59, 97, 80

⑥ Let's fill in each ☐ with a number.

① 100 and 15 make ☐.

② 100 and 8 make ☐.

⑦ Let's find the answers.

① 60 + 30

② 20 + 80

③ 90 − 60

④ 100 − 20

⑧ Let's find the answers.

① 34 + 5   ② 22 + 6

③ 7 + 51   ④ 8 + 40

⑤ 69 − 6   ⑥ 87 − 4

⑦ 57 − 7   ⑧ 95 − 5

⑮ Time (2)

pp.48～50

1   What time is it?

①

②

③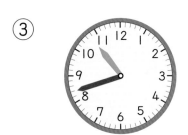

④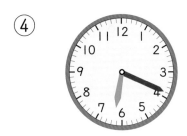

2   Let's draw the long hand on the clock.

①   2:50

②   7:25

③   11:33

④   1:46

⑯ Addition or Subtraction? Let's think by using diagrams.

pp.51～57

1 On a volunteer activity, Rei picked up 8 cans and Yuri picked up 6 more cans than Rei.

How many cans did Yuri pick up?

2 There are 11 apples. Pears are 7 fewer than apples.

How many pears are there?

3 The children are walking in line. Sayumi is the 5th child from the front. There are 3 children behind her.

How many children are there altogether?

⑱ Shapes (2)

pp.62～64

1 How many pieces of shape Ⓐ are needed to make the following shapes?

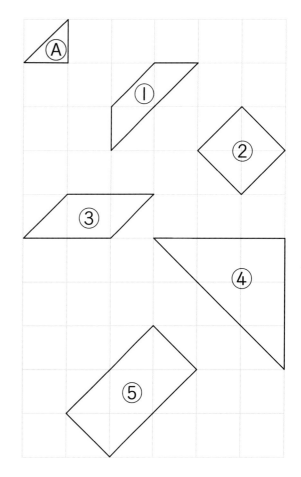

Clock ▼will be used in pages 48, 49, and 50.

Colored Pieces

▼ will be used in pages 62 and 63.

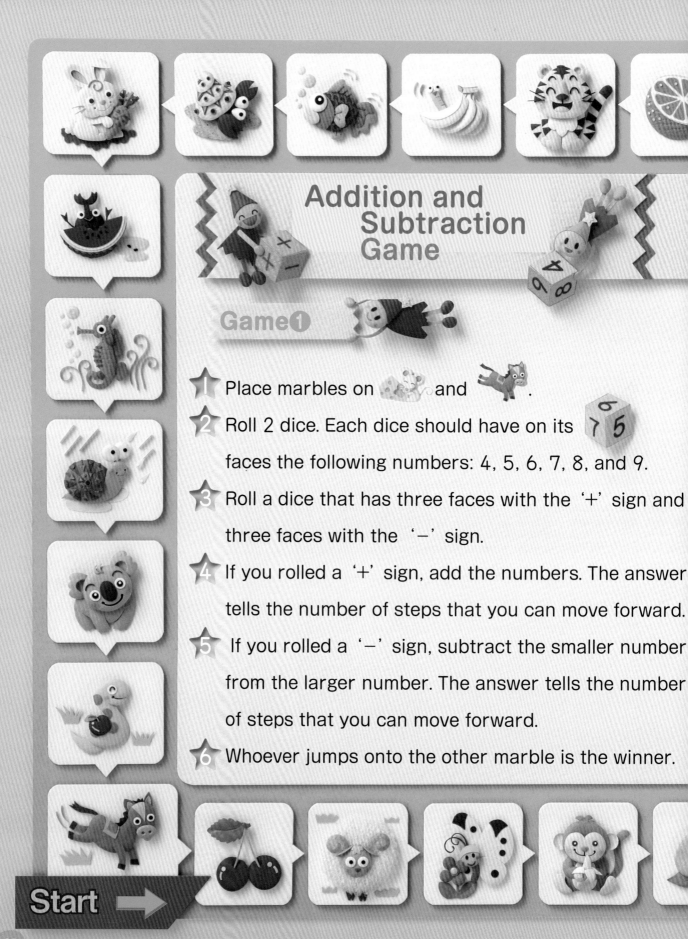

Addition and Subtraction Game

Game①

1. Place marbles on and .
2. Roll 2 dice. Each dice should have on its faces the following numbers: 4, 5, 6, 7, 8, and 9.
3. Roll a dice that has three faces with the '+' sign and three faces with the '−' sign.
4. If you rolled a '+' sign, add the numbers. The answer tells the number of steps that you can move forward.
5. If you rolled a '−' sign, subtract the smaller number from the larger number. The answer tells the number of steps that you can move forward.
6. Whoever jumps onto the other marble is the winner.

Start ➡

The player who jumps onto the other marble is the winner.

Game❷

⭐ Place marbles on 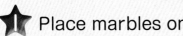 and  .

⭐ Throw a dice with the numbers 6, 6, 7, 7, 8, and 9 on its faces and another dice with the numbers 10, 11, 12, 13, 14, and 15 on its faces.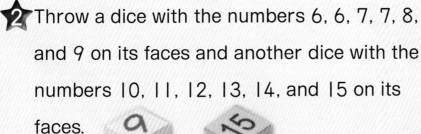

⭐ Subtract the smaller number from the larger number. The answer tells the number of steps that you can move forward.

⭐ Whoever jumps onto the other marble is the winner.

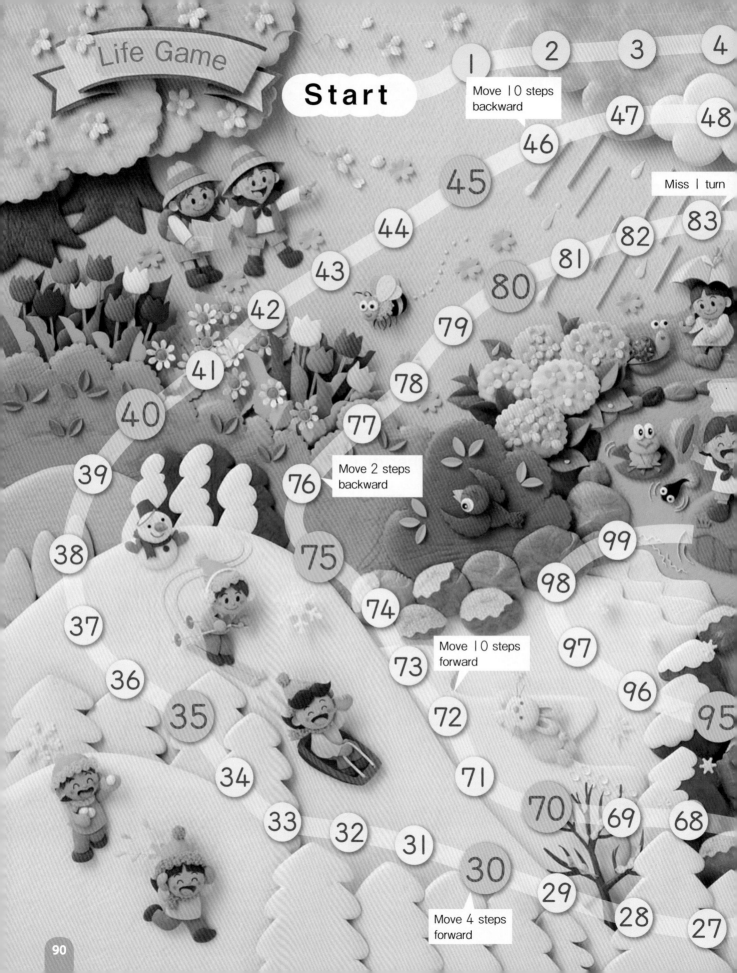

Editorial for English Edition:

Study with Your Friends, Mathematics for Elementary School

1st Grade, Vol.2, Gakko Tosho Co.,Ltd., Tokyo, Japan [2020]